Megan SMITH

BY ROBIN KOONTZ

ILLUSTRATED BY ELENA BIA

Rourke
Educational Media
rourkeeducationalmedia.com

A Division of
Carson
Dellosa
Education

WOMEN IN SCIENCE & TECHNOLOGY

Before Reading: *Building Background Knowledge and Vocabulary*

Building background knowledge can help children process new information and build upon what they already know. Before reading a book, it is important to tap into what children already know about the topic. This will help them develop their vocabulary and increase their reading comprehension.

Questions and Activities to Build Background Knowledge:

1. Look at the front cover of the book and read the title. What do you think this book will be about?
2. What do you already know about this topic?
3. Take a book walk and skim the pages. Look at the table of contents, photographs, captions, and bold words. Did these text features give you any information or predictions about what you will read in this book?

Vocabulary: *Vocabulary Is Key to Reading Comprehension*

Use the following directions to prompt a conversation about each word.

- Read the vocabulary words.
- What comes to mind when you see each word?
- What do you think each word means?

Vocabulary Words:
- diverse
- economy
- engineer
- influential
- innovation
- programmer
- solar-powered
- start-up

During Reading: *Reading for Meaning and Understanding*

To achieve deep comprehension of a book, children are encouraged to use close reading strategies. During reading, it is important to have children stop and make connections. These connections result in deeper analysis and understanding of a book.

 Close Reading a Text

During reading, have children stop and talk about the following:

- Any confusing parts
- Any unknown words
- Text to text, text to self, text to world connections
- The main idea in each chapter or heading

Encourage children to use context clues to determine the meaning of any unknown words. These strategies will help children learn to analyze the text more thoroughly as they read.

When you are finished reading this book, turn to the next-to-last page for **Text-Dependent Questions** and an **Extension Activity**.

TABLE OF CONTENTS

EVERYTHING IS INTERESTING!

Megan Smith was born curious about everything around her. She loved to create things. When Megan was four, she built a scarecrow out of sticks and tin foil.

Megan also loved learning how things worked. She even took apart her stepmom's bicycle. She wanted to figure out how it operated. She left the parts in a bucket.

At Megan's elementary school in Buffalo, New York, everyone had to enter the science fair. Megan was delighted. An energy crisis in 1974 gave her an idea. People should switch from fossil fuel to green energy! Megan used her father's tools to build a **solar-powered** house. She entered it in the science fair. That's when Megan realized she could create something important.

"If I hadn't had incredible teachers who made us do this science and tech stuff ... I wouldn't have known about this secret world of innovation," Megan said.

BUSY BEGINNINGS

Megan told her family that she wanted to become an **engineer**. Her grandfather didn't understand why a girl would want that. He was an engineer himself, but he wasn't used to seeing women in the job.

After high school, Megan attended the Massachusetts Institute of Technology (MIT). She worked on space exploration and other engineering projects. Megan was also part of a team that designed and built a solar-powered car. They raced it 2,000 miles (3,219 kilometers) across the Australian outback.

When she graduated from MIT, her grandfather was proud. He bragged about her to all his friends.

Megan's first jobs were with **start-up** companies, including General Magic. There, she helped develop some of the first smartphones.

Megan joined Google in 2003. She later became its vice president of business development. She helped Google celebrate the birthdays of **influential** women with Google Doodles. Her teams supported many new youth outreach programs, including a Google Science Fair.

"We need to know that women have always done these jobs ... even if they've been written out of the stories," Megan said.

Girls Around the World
In 2012, Megan cofounded the Malala Fund, which supports expanding education opportunities for girls around the world.

SERVICE TO THE COUNTRY

In 2014, Megan became the third United States Chief Technology Officer (US CTO). She was the first woman in the job. She advised U.S. president Barack Obama on bringing technology, **innovation**, and data to the American people. Megan and her teams updated the White House's old technology. They brought more technical talent with **diverse** backgrounds into the government. New tech jobs were created.

Megan designed a website for the White House. It was devoted to "the untold history of women in science and technology." The website features women such as Ada Lovelace, the world's first **programmer**. Encouraging women to work in technology became a big part of Megan's life.

Ada Lovelace

Grace Hopper

Maria Klawe

"See yourself in the field. Women have always been doing it, just not getting recognition or their stories told," Megan said.

Megan left the White House in 2017. Then she got busier than ever! She helped launch the Tech Jobs Tour. The tour goes to areas of smaller, more diverse populations. They want people to understand that technology is key to their future **economy** and community. The tour brings in people with different talents. They match them with companies who need their skills.

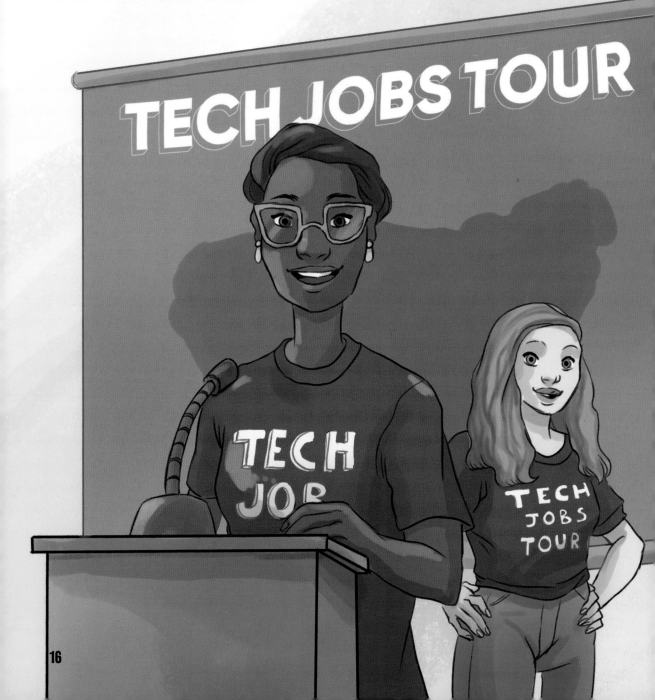

TECH JOBS TOUR

New York

Portland

Memphis

Indianapolis

Austin

Orlando

San Francisco

Megan works with students to encourage interest in science, technology, engineering, and math. She wants us all to work together to solve problems. She thinks almost anyone can be happy working in technology. They just need the chance to try it.

Solutions Summit

Megan is also the founder and chief executive officer (CEO) of Shift7. Shift7 helped organize the 2018 United Nations (UN) Solutions Summit. The summit unites people developing fresh ideas that could solve world challenges.

"There's never been a better time ... we have the ability to together solve some of the greatest challenges our world faces. So many of those solutions will be from scientific discoveries and technological innovations," Megan said.

TIME LINE

1964: Megan Smith is born on October 21 in Buffalo, New York.

1982: Megan graduates from City Honors School.

1986: Megan earns a bachelor's degree from Massachusetts Institute of Technology (MIT).

1987: Megan serves on the Solar Car Team and participates in the first Cross-Continental Solar Car race across the Australian outback.

1988: Megan earns a master's degree in mechanical engineering from MIT.

1996: Megan becomes COO of PlanetOut, an interactive company dedicated to the LGBT community.

1999: Megan marries Kara Swisher.

2001: Megan becomes CEO of PlanetOut.

2003: Megan joins Google Inc.

2012: Megan becomes a vice president of Google X.

2012: Megan cofounds the Malala Fund, which supports expanding educational opportunities for girls around the world.

2014: Megan becomes the first female United States Chief Technology Officer (US CTO).

2017: Megan founds Shift7.

2017: Megan is elected to the National Academy of Engineering.

2017: Megan helps launch the Tech Jobs Tour.

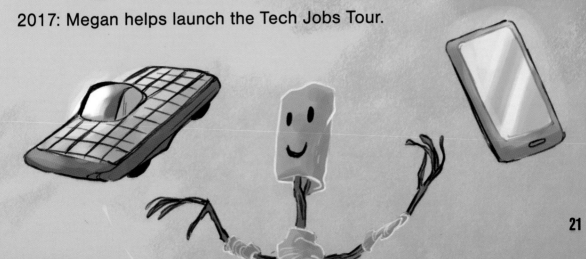

GLOSSARY

diverse (dye-VURS): varied or assorted

economy (i-KON-uh-mee): the way a country runs its industry, trade, and finance

engineer (en-juh-NEER): someone who is trained to design and build machines, vehicles, bridges, buildings, roads, or other technologies

influential (in-floo-EN-shuhl): having the power to affect or change someone or something

innovation (in-uh-VAY-shuhn): a new idea or invention

programmer (PROH-gram-ur): someone whose job it is to program a computer

solar-powered (SOH-lur-POU-urd): powered by energy from the sun

start-up (start-uhp): a newly established business

INDEX

TEXT-DEPENDENT QUESTIONS

1. What did Megan do to her stepmother's bicycle?

2. Where did Megan go to college?

3. What is the Malala Fund?

4. Who was President of the United States while Megan worked at the White House?

5. Why did Megan help create the Tech Jobs Tour?

EXTENSION ACTIVITY

Use the library or the internet to find three women in history who are famous for their contributions to technology. Write a sentence about each one that tells how their work helped make something better.

ABOUT THE AUTHOR

Robin Koontz loves to learn and write about everything from aardvarks to ziggurats. Raised in Maryland and Alabama, Robin now lives with her husband in the Coast Range of western Oregon. She enjoys figuring out how to repurpose things into useful contraptions. You can learn more on her blog: robinkoontz.wordpress.com.

ABOUT THE ILLUSTRATOR

Elena Bia was born in a little town in northern Italy, near the Alps. In her free time, she puts her heart into personal comics. She also loves walking on the beach and walking through the woods. For her, flowers are the most beautiful form of life.

www.rourkeeducationalmedia.com

Edited by: Kim Thompson
Cover and interior design by: Rhea Magaro-Wallace

Library of Congress PCN Data

Megan Smith / Robin Koontz
(Women in Science and Technology)
  ISBN 978-1-73161-429-2 (hard cover)
  ISBN 978-1-73161-224-3 (soft cover)
  ISBN 978-1-73161-534-3 (e-Book)
  ISBN 978-1-73161-639-5 (ePub)
Library of Congress Control Number: 2019932135

Rourke Educational Media
Printed in the United States of America,
North Mankato, Minnesota

Quote Sources:
Communications of the ACM, June 2015, Vol. 58 No. 6, Pages 39-43: An Interview with U.S. Chief Technology Officer Megan Smith by Vinton G. Cerf https://cacm.acm.org/magazines/2015/6/187315-an-interview-with-u-s-chief-technology-officer-megan-smith/fulltext;

Megan Smith on her background in engineering, the importance of innovation, and how we can diversify technology: https://www.makers.com/profiles/591f25476c3f64632d4fb856;

U.S. CTO Megan Smith: Four ways to get girls into STEM | Fortune Magazine, Published on Oct 13, 2015, https://www.youtube.com/watch?v=H2Y2mjmPwMs,

Huffington Post: THE BLOG 10/11/2013 05:23 pm ET Updated Jan 23, 2014 'Passion, Adventure and Heroic Engineering'... and Talent Inclusion By Megan Smith, https://www.huffingtonpost.com/megan-smith/women-in-tech_b_4086332.html

Megan Smith

Megan Smith thinks anyone would be happy working in technology if they have the chance to try it! The former Chief Technology Officer of the United States shows how fun the field of technology can be, and how people of all backgrounds can adventure together to solve the world's problems.

Alignment

This book supports the C3 Framework for Social Studies State Standards. Readers will learn about women who have shaped historical changes in the fields of science and technology.

Books in the series *Women in Science and Technology* include:

Annie Easley

Elizabeth Blackwell

Grace Hopper

Katherine Johnson

Mae C. Jemison

Megan Smith

Guided Reading Level: **P**

ISBN: 978-1-73161-224-3

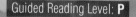

Rourke
Educational Media
rourkeeducationalmedia.co

A Division
Carson
Dellos
Educatio